Electricity and Magnetism

written by Maria Gordon
and
illustrated by Mike Gordon

Wayland

Simple Science

Series Editor: Catherine Baxter
Advice given by Audrey Randall – member of the Science Working Group
for the National Curriculum.

First published in 1995 by
Wayland (Publishers) Ltd
61 Western Road, Hove
East Sussex, BN3 1JD, England

British Library Cataloguing in Publication Data
Gordon, Maria
  Electricity and Magnetism. – (Simple Science Series)
  I. Title II. Gordon, Mike III. Series
  537

ISBN 0 7502 1599 2

Typeset by MacGuru
Printed and bound in Italy by G. Canale and C.S.p.A, Turin, Italy

Contents

Electricity is a sort of energy. It is called energy because it makes things happen. You cannot see electricity, but you can tell where it is...

...and where it isn't! It can make lights shine, fires hot and machines work.

Electricity has always been in the world...

it makes lightning...

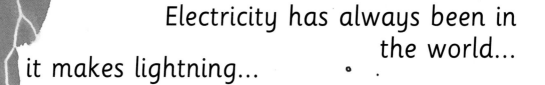

some animals hurt other animals with it...

it can make brushed hair stand on end...

It even helps hearts to beat!

6

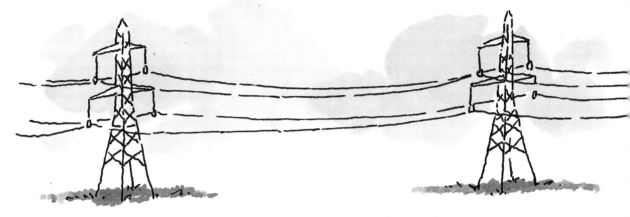

People have learnt how to make electricity. It is sent along wires into buildings. It goes into sockets, then plugs and along more wires into machines and other things which use electricity.

Electricity comes out of batteries, too.

Keep safe! Never play with sockets, plugs, wires or batteries. Electricity helps us, but it can hurt us, too!

Long ago, a Greek scientist called Thales rubbed a piece of stone with a silk cloth. This made straw and feathers stick to the stone. In English the stone is called amber. But the stone's Greek name is electron.

Hundreds of years later an English scientist called William Gilbert rubbed other things with silk, wool and fur. Paper and straw stuck to them, just like they stuck to amber. He could tell something was pulling the straw and paper. He called it electricity, like electron, amber's Greek name.

You can make
electricity too. Make
sure your hair is clean
and dry. Move a comb
through it quickly a
few times.

Hold the comb over little bits
of silver foil. Electricity pulls some
of the bits to the comb.

The comb makes static electricity. Static means something doesn't move. Static electricity stays in one place. It is made when some things rub together.

Electricity can move along things, too. When this happens it is called an electric current. An electric current needs a path, such as wire, to flow along.

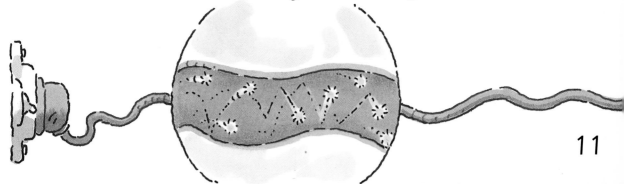

The path an electric current takes is called a circuit. This is because it must go back to where it started, like a circle.

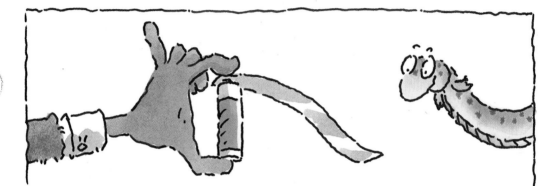

Ask a grown-up to help. Cut a strip of silver foil 10 cm by half a cm. Look at the picture. Hold the strip and an AA battery like this. Nothing happens! This is because the strip does not lead back to the battery.

Now hold the battery with one end of the foil touching each end. Wait until the strip begins to feel hot, then put the battery down. The foil feels hot because you have made a circuit and an electric current is flowing round it.

Things that let electricity move through them are called conductors.

Insulators are things which do not carry electricity. Look at the wires leading to things that use electricity. They are covered in insulators. This stops the electricity from hurting anyone.

You can find out if something is a conductor or an insulator.

Ask a grown-up to help. Cut two strips of silver foil 15 cm by 1 cm. Tape one end of each strip to each end of an AA battery.

Wrap the other end of one strip around the bottom of a torch bulb. Hold it there with a clothes peg.

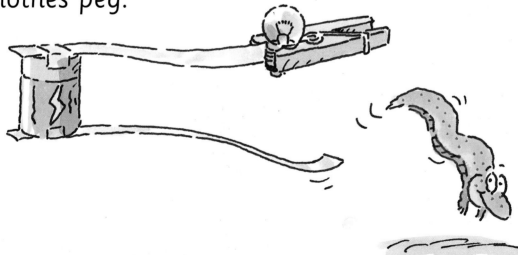

Move the loose end of the other strip so it is underneath the bulb. This makes a circuit. The electricity flows. It makes the bulb glow.

Now put paper between the foil and the bulb. Paper is an insulator. It does not make a bridge for electricity to move to the other strip. The bulb does not glow. The bulb glows if a coin is used instead of paper.

Why is this?

When electricity flows around iron or steel it makes them magnetic. This means they can pull other iron or steel things towards them. This is called magnetism. Magnetism is a sort of energy. You cannot see it, but you can watch it make things happen.

Ask a grown-up to help you.

Use a piece of plastic covered wire about 45 cm long.

Wrap it tightly round a steel nail. This makes a wire coil.

Leave 6 cm of wire at each end. Strip the plastic off the last inch of each end.

Tape both ends to a small battery.

Now the nail will pick up a paperclip. You have turned it into an electromagnet.

Hold the magnetized nail very
near to small pieces of...

plastic...

paper...

silver foil...

wood.

It does not pull these things towards it.
They cannot be magnetized.

Make the nail pick up a paperclip again.
Now lift one end of the wire off the battery.
This breaks the circuit. The electricity stops
flowing. The nail stops being an
electromagnet. It drops the paperclip.

Magnets are helpful because they only pick up steel or iron. People can use them to do things like... sorting out junk for recycling... picking up pins... holding doors closed.

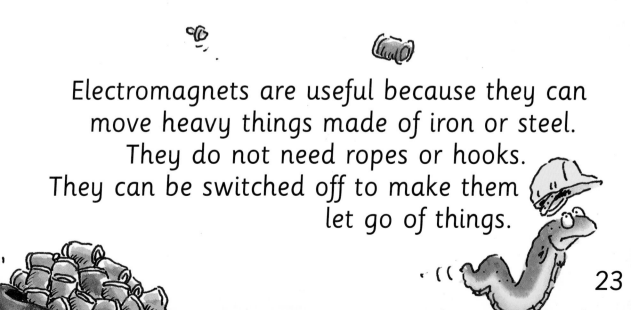

Electromagnets are useful because they can
move heavy things made of iron or steel.
They do not need ropes or hooks.
They can be switched off to make them
let go of things.

Use a magnet to pick up paperclips.
The ends pick up more.
The ends of a magnet
are called poles.

You can feel the power of poles.
Move the poles of two magnets together.
Two ends pull together.

The other two push apart.

Ask an adult to help.

Pour honey into a
small plastic tub until
it is about 1 cm deep.

Add half a teaspoon
of iron filings.
Mix them well.

Put the tub on top of a
magnet. The magnet pulls
the iron filings around it.
They move very slowly
through the honey.
The filings make a pattern.

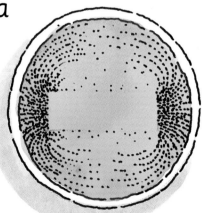

This shows the way the magnet pulls.

Magnets can help
make electricity, too.
Electricity will flow in
a wire coil if a magnet
is pushed in and out
of it very quickly.
A coil turning round a
magnet also makes
electricity needed for
houses and factories.
But something is
always needed to push
the coils or magnets.

Coal is often burnt to heat water. This makes steam which pushes giant magnets or huge coils. This uses up a lot of coal and makes the air dirty when the coal is burnt. Even batteries are filled with dangerous liquids.

It is important to save electricity.
This helps to keep the world clean
and safe and does not use up
all the things needed to
make electricity.

Now there are some things which make
electricity when light shines on them. Some
can use sunlight which won't run out or
make a mess. Which ones can you see here?

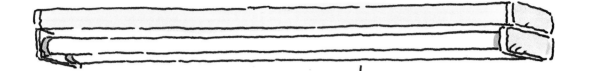

Many bulbs and machines are made so they do not use very much electricity. Check if you have some of these at home and school.

Electricity is saved by turning off machines, lights and heaters.

How can you help save electricity?

Notes for adults

The 'Simple Science' series helps children to reach Key Stage 1: Attainment Targets 1-4 of the Science National Curriculum.

Below are some suggestions to help complement and extend the learning in this book.

4/5   Write a story about the pictures.

6/7   Look at lightning conductors and pictures of electric eels, rays and torpedo fish. Borrow a pacemaker; talk to a user. Arrange a visit, safety talk and equipment demonstration by an ambulance crew. Make safety posters. Show battery corrosion. Visit construction sites revealing internal wiring.

8/9   Talk to people old enough to remember pre-electric times. Display pre-electric appliances. Investigate countries/communities without major electrification.

10/11   Rub balloons to make them stick to walls and ceilings. Write poems about lightning.

12/13   Make skill loop buzzers and light-up quiz boards. Investigate conversion to heat, sound and light. Trace telephone wires into receiver. Play electric synthesizers, guitars, etc.

14/15   Look at circuit testers. Strip off insulation from unused wire. Open up unused plugs. Display the safety gear used by electricians. Explore battery powered circuit paths inside torches, etc.

16/17   Investigate fluorescent lighting – electricity making gases glow. Talk about dangers of appliances near water. Investigate Franklin, Galvani and Volta. Note eighteenth-century lack of women scientists.

18/19   Make a fridge display of items to do with electricity using magnets.

20/21	Make a fishing game with magnets on lines. Use paperclipped paper fish and 'control' paper fish without paperclips.
22/23	Investigate the history of magnets, from Magnesia to Chinese chariot compasses, etc. Visit a car crushing plant.
24/25	Investigate the Earth as a giant magnet. Make floating needle compasses. Find toys that use magnets. Play 'people magnets': label chest and backs as poles; spread out, then move, joining together or apart as poles move near each other.
26/27	Visit a power station. Look at electricity meters and bills; make graphs showing seasonal variation. Display motors showing coils, etc.
28/29	Investigate solar-powered cars, satellites, etc. Arrange a photo-electric cell-based security system demonstration. Spend a day without using electricity!

Other books to read

Batteries, Bulbs and Wires by David Glover (Kingfisher Books, Grisewood and Dempsey, 1993)
Electricity by John Williams (Wayland, 1991)
My First Batteries and Magnets Book by Jack Challoner (Dorling Kindersley, 1992)
Physics for Every Kid by Janice Van Cleave (John Wiley, 1991)

Index